the SECRET signs of NATURE

written by

CRAIG CAUDILL

illustrated by

CARRIE SHRYOCK

MAGIC CAT PUBLISHING

NEW YORK

DISCLAIMER

THIS BOOK HAS BEEN MADE TO HELP CONNECT YOU WITH NATURE. IT'S NOT INTENDED TO BE A SURVIVAL HANDBOOK, SO PLEASE MAKE SURE YOU ARE ACCOMPANIED BY AN ADULT AS YOU VENTURE INTO THE GREAT OUTDOORS . . .

LEARNING TO READ THE SIGNS OF NATURE WILL CHANGE YOUR LIFE.

IT WILL TRANSFORM YOUR EXPERIENCE OF BEING OUTSIDE AND ENABLE YOU TO DECIPHER THE SECRET STORIES OF WILDLIFE.

THE SKILLS OF A NATURALIST AND NATURE NAVIGATOR WILL BECOME YOUR COMPASS, GPS, REMOTE CAMERA TRAP, PARABOLIC MICROPHONE, AND THERMAL-IMAGING CAMERA . . . ALL ROLLED INTO ONE! YOU CAN NEVER KNOW EVERYTHING ABOUT THESE TIMELESS, ENDLESS SKILLS—THERE IS ALWAYS MORE TO LEARN—BUT TO ME, THAT IS THE JOY. EVERY NEW BIRDSONG YOU DISCOVER, EVERY PRINT YOU LEARN TO RECOGNIZE, EVERY WILD SIGNPOST YOU CAN NOW READ . . . EVERY SINGLE ONE WILL MAKE YOUR WORLD RICHER.

Steve Backshall
Member of the Order of the British Empire
BAFTA-WINNING EXPLORER AND NATURALIST

The secret signs of . . .

RAINBOWS, CLOUDS, AND STORMS

OCEANS, PUDDLES, AND WATER

MAMMALS, INSECTS, AND BIRDS

TREES, FLOWERS, AND FUNGI

THE SUN, MOON, AND STARS

GETTING STARTED

RAINBOWS THAT FORECAST THE WEATHER, BIRDS THAT TELL THE TIME, PUDDLES THAT POINT SOUTH . . . IN THIS BOOK, YOU'LL FIND HUNDREDS OF TIPS FOR READING NATURE'S SECRET SIGNS, SOURCED FROM DECADES SPENT WALKING ON THE LAND NEAR MY HOME AND RESEARCH FROM AROUND THE WORLD. THESE SIGNS HAVE BEEN USED FOR MILLENNIA BY OUR ANCESTORS. UNTIL VERY RECENTLY IN OUR HISTORY, HUMANS TRAVELED WITHOUT MAPS, USING WAYS OF NAVIGATING THAT MOST OF US HAVE NOW LONG FORGOTTEN. THESE AREN'T SURVIVAL SKILLS, BUT A MEANS TO ENRICH JOURNEYS AND CONNECT WITH THE WORLD AROUND US. SO PORE OVER THESE PAGES, THEN GO OUTSIDE AND DISCOVER THE WONDERS OF NATURE FOR YOURSELF . . .

A YOUNG NAVIGATOR'S MOST IMPORTANT TOOLS

THIS BOOK WILL SHOW YOU HOW TO EXPLORE NATURE WITH THE MOST USEFUL TOOLS: YOUR SENSES. WHEN MOST OF US SPEND TIME OUTDOORS, A SINGLE SENSE—SIGHT—CAN DOMINATE OUR EXPERIENCE. BUT WE CAN USE MANY OF OUR SENSES TO "READ" EVERYTHING NATURE HAS TO OFFER . . .

LOOK

Take in details. What kinds of trees or plants are around you? What do their leaves look like? Look up at the sky, then at the path ahead, and finally look down at the ground. What tracks can you see?

LISTEN

Close your eyes. Listen closely to what is around you. Can you identify the birds in your area? What is happening in the spaces between sounds? Take a few steps and listen to the sound of your footsteps. What message do you hear?

SMELL

Take in a slow, deep breath through your nose, then exhale out through your mouth. Can you smell flowers? Or rain? Close your eyes and breathe in the scent again. What do you smell?

TOUCH

Select a tree or plant that is not poisonous to touch. Move your fingers to take in the details with your fingertips. Is it rough or smooth? What can you feel?

BEFORE YOU SET OFF . . .

THROUGHOUT THIS BOOK, YOU WILL COME ACROSS THIS COMPASS ICON. IT'S CALLED A COMPASS ROSE AND YOU'LL FIND IT ON PAGES WHERE WE'RE DISPLAYING HOW TO ORIENT YOURSELF OR FIND YOUR POSITION—THAT'S BECAUSE A COMPASS ROSE IS A SYMBOL THAT SHOWS CARDINAL DIRECTIONS.

CARDINAL DIRECTIONS

Cardinal directions are the four main directions of a compass—north, south, east and west—which are also known by their initial letters: N, S, E, and W. A cardinal direction can also be referred to by first using the word "due." So, for example, the cardinal direction of south can also be referred to as "due south."

PREVAILING WINDS

Prevailing winds are winds that blow from a single direction over a specific area. Generally, prevailing winds blow east-west rather than north-south. This happens because Earth's rotation causes what is known as the Coriolis effect. The Coriolis effect makes wind systems twist counterclockwise in the northern hemisphere and clockwise in the southern hemisphere.

BE CURIOUS

Pay careful attention to what is all around you . . .
Everything in nature has meaning—
the key to discovering those meanings is curiosity!

Now let's join these two young navigators as they discover the joy of finding their way using nature's clues.

You can never reach the end of a **RAINBOW**. It will always continue to move with you until it **DISAPPEARS**.

Rainbows appear in the part of the sky **OPPOSITE** the sun.

In the morning, the sun is in the **EAST**, so to see a rainbow, you must be facing toward rain in the **WEST**.

OBSERVE *a* RAINBOW *to decipher its* MESSAGE

RAINBOWS DON'T ONLY LOOK MAGICAL—THEY ARE USEFUL, TOO.

THE NEXT TIME IT RAINS, NOTE HOW SOON THE SUN COMES OUT. IF IT COMES OUT QUICKLY, ITS LIGHT WILL SHINE THROUGH THE RAIN DROPLETS IN THE AIR, THROWING A VIVID DISPLAY OF COLOR INTO THE SKY . . . A RAINBOW! BY DECIPHERING THE SECRET MESSAGES WITHIN A RAINBOW'S COLORFUL ARC—LIKE THIS ONE IN A COLORADO SKY—YOU CAN FIND THE CARDINAL DIRECTIONS, FORECAST THE WEATHER, AND EVEN TELL THE TIME.

HUMMINGBIRDS can see more colors of the rainbow than humans. Their eyes and brains have adapted to see **ULTRAVIOLET LIGHT**.

WEATHER LORE

RAINBOWS NEED TWO INGREDIENTS: SUNLIGHT AND WATER. WHEN LIGHT PASSES THROUGH WATER, IT BENDS AND SPLITS INTO A SPECTRUM OF LIGHT: A GROUP OF FAMILIAR COLORS THAT INCLUDE RED, ORANGE, YELLOW, GREEN, BLUE, INDIGO, AND VIOLET. IT'S THESE COLORS THAT CAN PROVIDE WEATHER SIGNALS . . . HAVE YOU HEARD THE PHRASE "RAINBOW IN THE MORNING GIVES YOU FAIR WARNING"? IT MEANS A RAINBOW IN THE MORNING INDICATES THAT RAIN IS IN THE NEAR FUTURE!

In the late afternoon, the sun is in the **WEST**, so after the rain has passed by, face toward the **EAST**, where you'll see a glorious rainbow.

Some of the largest rainbows occur just before **SUNSET**, so if you see a full **SEMICIRCLE**, you know it will soon be dusk.

With the weather coming in from the west in the northern hemisphere, a rainbow in the morning is usually a sign you're about to get **WET**, whereas one in the afternoon means the **SUN** will be out soon.

Upon sensing the vibrations of **RAIN** on the grass above, **EARTHWORMS** come out of their burrows to the surface. They find it much easier to travel across the soil when it's wet.

High winds can flatten the top of a **CUMULONIMBUS CLOUD** into an anvil-like shape (with a flat top and concave sides). A storm is usually **HEADING IN THE DIRECTION** that the anvil is **POINTING** toward.

A **DIMMING SKY** filled with **DARK, ROILING CLOUDS** is a signal to seek shelter.

When viewed from the ground, cumulonimbus clouds look ominous. They have **LOW, DARK BASES,** filled with **HEAVY WATER DROPLETS** . . . ready to fall on your head!

LOOK UP *at the clouds* *to* PREDICT A THUNDERSTORM

THE SKY DARKENS, THE WIND PICKS UP, AND CLOUDS FILL THE SKY . . .

THEN A LOW RUMBLE ANNOUNCES A THUNDERSTORM IS ON ITS WAY! BY LOOKING UP AT THE SKY, YOU CAN PREDICT WHETHER A THUNDERSTORM IS ON THE HORIZON. THERE ARE PLENTY OF OBSERVABLE CLOUD DETAILS: THE SIZE, SHAPE, DEPTH, AND COLOR OF THE CLOUD, AND EVEN HOW HIGH IT IS IN THE SKY. SO LET'S GO CLOUD SPOTTING IN MONTENEGRO'S DURMITOR NATIONAL PARK AND READ NATURE'S SIGNPOSTS . . .

HOW A CLOUD FORMS

CLOUDS FORM WHEN WARM AIR RISES FROM THE GROUND INTO THE ATMOSPHERE. AS IT COOLS DOWN, TINY MOLECULES OF WATER IN THE AIR CLUMP TOGETHER INTO LARGER WATER DROPLETS OR ICE CRYSTALS. AS MORE DROPLETS COLLECT, A CLOUD IS MADE. CUMULONIMBUS CLOUDS ARE ONE OF THE MOST RECOGNIZABLE FORMS, AND THEY GET TALLER AND TALLER UNTIL THEY BECOME HUGE POWERHOUSES, STORING ENERGY THAT PRODUCES A THUNDERSTORM. FORTUNATELY, A THUNDERSTORM OFTEN LASTS FOR ONLY AROUND THIRTY MINUTES, BUT THE RAINFALL IS VERY HEAVY, SO IT'S BEST TO KNOW WHEN TO SEEK SHELTER.

Clouds that signal an oncoming storm aren't always **BLACK**; they can take on **PURPLE**, **YELLOW**, or even **GREENISH HUES**.

To avoid **AIR PRESSURE** changes, birds—like the **CROSSBILL**—fly **CLOSER TO THE GROUND** when a storm is on the way.

If you are in water and a thunderstorm rolls in, return to shore immediately to find **SAFE SHELTER**.

FROGS can detect changes in **ATMOSPHERIC PRESSURE**, which is useful because rain creates optimal conditions for females to lay eggs. You can hear males' **MATING CALLS** just before rainfall.

On **PRECIPICES** high in the mountains, **SNOW STRIPS** appear along **SMALL, SHELTERED ROCK OUTCROPS**. The longer the strips, the stronger the **PREVAILING WIND**.

Snow **MELTS** more quickly on whichever side of a hill faces the **SUN**. In the northern hemisphere it is the **SOUTH-FACING** hillsides, and in the southern hemisphere it is the **NORTH-FACING** hillsides.

Look at the **BASE OF A TREE**; snow is **TIGHTLY PACKED** on the **WINDWARD SIDE**.

STUDY SNOW *to find your bearings*

SNOW MAKES ITS MARK ON THE ENVIRONMENT AROUND US.

WHEN THE TEMPERATURE IS BELOW FREEZING, WATER VAPOR IN THE AIR TURNS TO ICE. IT FORMS AROUND TINY PARTICLES OF DIRT THAT HAVE BEEN CARRIED BY THE WIND, CREATING ICE CRYSTALS. THESE ICE CRYSTALS STICK TO ONE ANOTHER, PRODUCING WHAT WE CALL SNOWFLAKES. WHEN ENOUGH ICE CRYSTALS STICK TOGETHER, THE SNOWFLAKES BECOME HEAVY AND FALL. IF THEY FALL THROUGH MOIST AIR, THEY STICK TOGETHER AND CREATE "WET SNOW" . . . PERFECT FOR ROLLING SNOWMEN AND CATCHING FLAKES ON OUR TONGUES! IT IS THIS WET SNOW THAT STICKS TO OBJECTS ON THE GROUND AND LEAVES BEHIND CLUES FOR US TO NAVIGATE FROM. TAKE A LOOK AROUND THIS MOUNTAIN VALLEY IN THE SWISS ALPS AND STUDY THE SIGNS LEFT BEHIND LONG AFTER SNOW HAS STOPPED FALLING . . .

If there is an ice field in the distance, the underside of **LOW CLOUDS** will appear **LIGHTER** than normal, a phenomenon known as an **ICEBLINK.**

Snow blows against trees in the valley, creating **VERTICAL STRIPS** on the sides of trees that are facing the **PREVAILING WIND.**

CHAMOIS have sharp, **POINTED TOES** on their **CLOVEN HOOVES,** which create unique imprints in the snow.

DRIFT SNOW

WIND TRANSPORTS SNOW AND IS THE CAUSE OF "DRIFT SNOW." THIS IS OFTEN FRESHLY-FALLEN SNOW THAT IS MOVED BY THE WIND BEFORE IT HITS THE GROUND, BUT STRONG WINDS CAN ALSO MOVE SNOW ON THE GROUND. SNOW SETTLES WHEN THE WIND HAS VIRTUALLY STOPPED, USUALLY AGAINST THE WINDWARD SIDE OF A LARGE, STATIONARY OBJECT, LIKE A TREE. SNOW CAN STICK TO THE SIDES OF TREES, CREATING "SNOW STRIPS," AND IT CAN PILE UP AT THE BASE.

MOUNTAIN HARES are indicators of the **SEASON**. They have brown coats for most of the year, but during winter, they develop **WHITE COATS** as **CAMOUFLAGE** from predators.

PREDICT *a* RAIN SHOWER *by* CONNECTING *with your* SENSES

CLOSE YOUR EYES, TAKE A DEEP BREATH IN, AND SMELL THE AIR.

HIGH PRESSURE AND HUMIDITY—CONDITIONS THAT OFTEN BRING RAIN—MAKE OUR ABILITY TO SMELL THINGS MUCH SHARPER. RAIN HAS NO SCENT, BUT MOMENTS BEFORE IT FALLS ONTO DRY GROUND, AN EARTHY SMELL CALLED PETRICHOR PERMEATES THE AIR. FLORA AND FAUNA ARE SENSITIVE TO IT . . . AS ARE YOU! CONNECT WITH YOUR SENSES AND FIND YOUR WAY AROUND THIS FOREST FLOOR IN FONTAINEBLEAU, FRANCE. IT SMELLS LIKE RAIN IS HEADING YOUR WAY—IT'S TIME TO GET YOUR RAINCOAT ON!

NOSE TO THE WIND

PETRICHOR IS CREATED FROM A PAIR OF CHEMICAL REACTIONS: SOME PLANTS SECRETE OILS DURING DRY PERIODS, AND WHEN IT RAINS, THESE OILS ARE RELEASED INTO THE AIR ALONG WITH CHEMICALS PRODUCED BY SOIL-DWELLING BACTERIA KNOWN AS ACTINOMYCETES. THESE COMPOUNDS COMBINE TO CREATE THE PETRICHOR SCENT WHEN RAIN HITS THE GROUND. IF THE RAINFALL IS HEAVY ENOUGH, THE PETRICHOR SCENT CAN TRAVEL DOWNWIND AND ALERT US THAT RAIN IS ON THE WAY. AFTER THE RAIN HAS PASSED, THE SCENT DISAPPEARS AND THE GROUND BEGINS TO DRY . . . BUT THE ACTINOMYCETES REMAIN LYING IN WAIT FOR WHEN IT MIGHT RAIN AGAIN.

Just before rainfall, some **PINECONES CLOSE UP** to protect their seeds from getting drenched.

The **RAIN'S SCENT** has a **CHEMICAL SIGNAL** that attracts insect-like **SPRINGTAILS.**

They eat the part of the **BACTERIA** that produces spores, so to these tiny creatures, the smell of the rain means it is dinnertime! In exchange for a meal, the little hairs on the springtails get covered in **SPORES** that they then spread around to colonize new habitats.

Sensing the **ATMOSPHERIC CHANGES, BEES** can predict rain is coming up to a day before it arrives. Watch them buzz about as they work harder and spend more time foraging for nectar.

Use the SHAPES OF DUNES *to determine the* WIND'S DIRECTION

AS THE WIND BLOWS, YOU'LL SEE SHIFTING SHAPES OF SANDY SCAPES.

A POWERFUL WIND CAN CARRY PARTICLES OF SAND VERY, VERY FAR BEFORE THEY DROP TO THE SURFACE. AS THEY ROLL, BOUNCE, AND FLY ACROSS THE GROUND, THE PARTICLES CREATE SMALL, WAVE-SHAPED RIPPLES, WHICH CAN BUILD UP INTO LARGER STRUCTURES CALLED SAND DUNES. SAND DUNES CAN GROW VERY HIGH AND VERY LONG—LIKE THOSE HERE IN THE KALAHARI DESERT, WHICH STRETCHES OVER BOTSWANA AND PARTS OF NAMIBIA AND SOUTH AFRICA. THEY CAN ALSO TAKE THE FORM OF VARIOUS SHAPES, SUCH AS CRESCENTS AND STARS. IT'S THESE UNIQUE SHAPES THAT REVEAL SECRET SIGNS IN THE DESERT . . .

The wide side of a **CRESCENT DUNE** is its **WINDWARD** side, with a small, semi-circular slip face on the other side.

A **CRESCENT DUNE** has two **"HORNS,"** the distance between which can reach over 300 feet from tip to tip.

In areas with **WIND BREAKS**, like large rocks, look closely at the **SAND PARTICLES**: They have different **COLORS**, **WEIGHTS**, and **SIZES** depending on the type of rock they originated from. Light particles will get blown past wind breaks and spread out on their **LEEWARD** side.

With their star-shaped tops, **STAR DUNES** are generally the tallest dunes and have three or more "arms," usually of irregular shapes.

They form by a wind that blows in **SEVERAL DIRECTIONS** from the same spot. As a result, it's ill-advised to use these dunes for orientation.

LINEAR DUNES are shaped by winds blowing in two directions and are aligned **PARALLEL** to the **STRONGEST WIND**.

These long, narrow sand dunes form in **RIDGES** that run roughly **SIDE BY SIDE**.

SHAPES OF DUNES

IF YOU ARE SEEKING DIRECTION, THE SHAPE OF SAND DUNES WILL BE YOUR FRIEND. THEIR SHAPES ARE EASY TO RECOGNIZE BY THEIR USEFUL NAMES: CRESCENT, LINEAR, AND STAR. EVERY DUNE HAS A WINDWARD SIDE AND A SLIP FACE. A DUNE'S WINDWARD SIDE IS THE SIDE THE WIND IS BLOWING AGAINST. A DUNE'S SLIP FACE IS SIMPLY THE SIDE WITHOUT WIND. HOW CAN YOU TELL THE DIFFERENCE? THE SLIP FACE SIDE IS USUALLY A LOT SMOOTHER.

There is little to break the winds here, so vegetation will **LEAN AWAY** from the **PREVAILING WIND**, as it's consistently pushed from the constant current.

TRAIL *the* SWELL *of the* SEA

THE REGULAR RHYTHM OF THE SEA'S SWELL CAN TAKE YOU TO SHORE.

OUT AT SEA, YOU'LL FIND GREAT BODIES OF PULSING WATER KNOWN AS SWELLS. THESE ARE CREATED BY FAR-OFF WINDS PUMMELING THE WATER AND PASSING ON THEIR ENERGY, WHICH TRAVELS THROUGH THE SEA IN A SERIES OF REGULAR WAVES. A SWELL CAN TRAVEL ENORMOUS DISTANCES, BUT IF YOU ARE CLOSE ENOUGH TO LAND, FOLLOWING THE SWELL WILL GENERALLY LEAD YOU TOWARD THE SHORE . . . SOMETHING THAT NATURE'S SURFERS—THE SEA LIONS ON THE GALÁPAGOS ISLANDS IN THE PACIFIC OCEAN—HAVE LEARNED TO MAKE THE MOST OF AS THEY RIDE BACK TO THE BEACH!

Moving in a **RHYTHM**, the **SWELL** journeys to the **SHORE**. Before modern technology, sea captains could recognize these rhythms and know what to expect in different areas of the ocean.

What begins as ripples on the surface grows into a **POWERFUL SWELL** as more and more **WIND ENERGY** is added.

Wind causes water to move and make tiny waves called **RIPPLES**. Ripples are an indicator of the direction of the **PREVAILING WIND**.

SEA LIONS glide gracefully through the **SWELL** that surges them to shore and safely back to their **COLONY**.

It's sunrise, and these **TERNS** are heading out to sea to hunt fish. When the sun sets, they will return to rest. By following their journey, you can position yourself toward **OPEN WATER** or toward **LAND**.

WIND WAVES

WIND WAVES ARE A DISPLAY OF THE ENERGY THAT HAS BEEN TRANSFERRED TO THE SEA FROM THE WIND. THIS ENERGY TRAVELS DOWNWIND AWAY FROM THE SOURCE, LIKE A POND'S RIPPLES WHEN A STONE IS DROPPED IN. THESE WAVES THAT MOVE AWAY FROM THEIR SOURCE ARE KNOWN AS SWELLS. THE STRONGER THE WINDS, THE BIGGER THE SWELL WILL BE AND THE FARTHER IT WILL TRAVEL. THE LONGER THE WIND BLOWS, THE LONGER THE SWELL WILL PERSIST, EVEN AFTER THE WIND HAS STOPPED OR CHANGED DIRECTION.

Swells can **FORECAST THE WEATHER**, with a large **SWELL** telling you that a **ROUGH STORM** is coming and it's time to head back to land.

Large **MARINE IGUANAS** can swim through **SWELLS**, **DIVING** to a depth of well over **30 FEET** to munch on green algae.

DECODE *the color of the* OCEAN

THE ARRAY OF VIVID BLUES, GREENS, AND BROWNS SPEAK TO US . . .

OUR OCEANS HAVE NO STREET SIGNS, AND THEY DO NOT EASILY GIVE AWAY THE SHAPES OF THE BOTTOM BENEATH THEIR SURFACE. FOR THOUSANDS OF YEARS, SAILORS HAVE INSTEAD USED THE COLOR OF THE SEA TO HELP THEM NAVIGATE. THE COLOR OF SEAWATER IS DETERMINED BY WHAT'S UNDER IT AND WHAT'S IN IT—AND BY THE COLOR OF THE SKY. AS SUNLIGHT PIERCES THE WATER, IT STRIKES THE BOTTOM AND, DEPENDING ON WHAT'S DOWN THERE, IT PRODUCES A TELLTALE COLOR . . . HEAD OUT TO SOME OF THE WORLD'S CLEAREST WATER IN THE BAHAMAS TO SEE WHAT THE COLORS OF THE OCEAN TELL US.

WAVES get **STEEPER** and **BREAK** when the water becomes too shallow to support them. If you see a spot where the water is heaping up, you know it's shallower there.

Birds like the **WHITE-TAILED TROPIC BIRD** can't stand on water. Keep an eye out for any that are "standing" and avoid the area as there may be a **ROCK** just below the water's surface.

As a general rule, the **DEEPER** the **SHADE OF BLUE**, the **DEEPER THE WATER**.

If water appears white or pale, it's most likely **CLEAR, SHALLOW WATER** lying over **WHITE SAND**.

BOTTLENOSE DOLPHINS emit a series of **HIGH-PITCHED** clicks, then listen for the echoes to bounce off their surroundings. Based on these echoes, they can determine the size and shape of **NEARBY OBJECTS**.

SHALLOW WATER is typically much easier to identify when the sun is not too low in the sky and when it's **BEHIND** you, preferably over your shoulder.

SEA RHYMES

SAILORS CAME UP WITH RHYMES TO DECODE WHAT THE COLOR OF THE WATER TOLD THEM WHILE SAILING OUT AT SEA.

"BROWN, BROWN, RUN AGROUND" MEANS THE WATER IS NOT DEEP, SO IT'S BEST AVOIDED.

"WHITE, WHITE, YOU JUST MIGHT" MEANS A SANDY BOTTOM THAT MAY BE SHALLOW, BUT IT DEPENDS ON THE WATER'S CLARITY.

"GREEN, GREEN, IN BETWEEN" MEANS THE WATER IS STARTING TO GET SHALLOW AND IT'S MOSTLY SAFE.

"BLUE, BLUE, SAIL ON THROUGH" MEANS THE WATER IS DEEP ENOUGH TO GO THROUGH

If you spot a **BROWN PATCH**, it usually means there is something—like a **CORAL HEAD**—close to the surface.

GREEN WATER shows that the water is in between depths; sometimes it means there is **SEAGRASS** on the bottom (and the color is being reflected back to you), but other times, it can be an indication of **SHALLOW WATER**.

Explore a COASTLINE *using* SEAWEED

YOU MIGHT NOT KNOW IT, BUT SEAWEED PLAYS A MAJOR ROLE IN MARINE ECOSYSTEMS.

SEAWEED CREATES HABITATS THAT PROVIDE SHELTER AND FOOD FOR THOUSANDS OF MARINE CREATURES, AND IT CAN BE FOUND ALIVE IN THE OCEAN OR WASHED UP ON THE SHORE. WHEN GROWING OUT AT SEA, MOST SEAWEED VARIETIES WILL ATTACH THEMSELVES TO ROCKS TO KEEP THEM SAFELY IN ONE PLACE. HOWEVER, STORMS CAN BREAK OFF PIECES OF SEAWEED AND THIS IS OFTEN WHAT WE THEN FIND WASHED UP. (OCCASIONALLY, WE CAN FIND PIECES WITH THEIR "ANCHOR" STILL ATTACHED!) RANGING IN COLOR FROM RED TO GREEN, AN AMAZING ARRAY OF SEAWEED CAN BE FOUND ALONG OUR COASTS AND CAN TELL US KEY INFORMATION ABOUT OUR CLIMATE. JUST BY WALKING ALONG THE COASTLINE—LIKE HERE IN HOKKAIDO, JAPAN—YOU CAN SPOT MANY SIGNS THAT SHOW JUST HOW VITAL SEAWEED IS.

When a **LIGHT WIND BLOWS** across the surface of calm water, it can create a dark area of **BIGGER RIPPLES**. This pattern is called the **CAT'S-PAW** and gives us a map of where the strong gusts of wind are.

SEA OTTERS sometimes float in forests of kelp, or **GIANT SEAWEED**, in which they entangle themselves to provide **ANCHORAGE** in the swirling sea.

KELP, a large brown seaweed, depends on **SUNLIGHT** for its survival. This means it will grow only in **SHALLOW WATER** where sunlight can reach the seabed—from the **LOW-TIDE MARK** down to **200 FEET** if the water is clear.

TUFTED PUFFINS create **BURROWS** in **ROCKY CLIFFS** and build nests lined with grasses and **SEAWEED**.

WHAT IS SEAWEED?

ALTHOUGH IT MAY LOOK LIKE IT, A SEAWEED ISN'T ACTUALLY A PLANT . . . ONE OF THE BIG DIFFERENCES IS THE LACK OF ROOTS. INSTEAD OF RECEIVING WATER FROM THE GROUND, SEAWEED ABSORBS NUTRIENTS FROM THE WATER THAT IS ALL AROUND IT. HOWEVER, SEAWEED DOES PHOTOSYNTHESIZE (USE SUNLIGHT TO MAKE FOOD), ALTHOUGH NOT ALWAYS USING CHLOROPHYLL (THE SUBSTANCE THAT MAKES PLANTS GREEN). SEAWEED IS ACTUALLY ALGAE; A COVER-ALL TERM FOR RELATIVELY SIMPLE PHOTOSYNTHESIZING, PRIMARILY MARINE ORGANISMS. THEY ARE NOT PLANTS, NOT ANIMALS, AND NOT FUNGI, BUT SOMETHING ELSE.

A piece of **SEAWEED** hung up will become **DAMP** in **ADVANCE OF RAIN** as seaweed absorbs water in the atmosphere. However, while this is often caused by **HUMID AIR**, it can also happen before **GOOD WEATHER** when **DEW** is forming.

BROWN SEAWEED is common in **COLD WATER** and are found along **ROCKY SHORES**.

Most **GREEN SEAWEED** occurs in fresh water, but one marine species, called **SEA LETTUCE**, can be found **ATTACHED** to **ROCKS** (or living in **ROCK POOLS** if it has become detached!).

LOCATE WATER *in the* DESERT

WATER IS VITAL TO LIFE.

BUT IN A HOT, ARID DESERT LIKE THE NAMIB IN SOUTHERN AFRICA, FINDING WATER CAN BE A COMPLICATED TASK. RAIN IS THE MAIN SOURCE OF WATER HERE AND ONCE THE BRIEF SEASONAL RAINS STOP, WILDLIFE FLOCK TO MEAGER WATERING HOLES THAT HAVE BEEN LEFT BEHIND. OFTEN FOUND ON LOWER GROUND, THESE HOLES ATTRACT LARGE BEASTS SUCH AS AFRICAN BUSH ELEPHANTS, WHILE SMALL INSECTS SUCH AS DRAGONFLIES FLY AROUND TO BREED HERE. UNDERSTANDING THE MOVEMENTS OF DESERT CREATURES AND WATCHING PLANT LIFE WILL HELP YOU SPOT THE SIGNS THAT WATER IS NEARBY.

With their **TRUNKS MOVING BACK AND FORTH**, the **AFRICAN BUSH ELEPHANTS'** sense of smell is in constant use, detecting new scents and information, and smelling water up to **12 MILES** away.

TRACKS on the ground from more than **ONE SPECIES** heading in the **SAME DIRECTION** are a common sign of water ahead.

CAPE TURTLE DOVES can live only near **FRESH WATER**. They tend to drink before 10 A.M. and after 4 P.M., **FLYING LOW** back to their nests once they have had their fill.

SPOTTING FLIES and **MOSQUITOES** together is a giveaway that a **WATER SOURCE** is nearby.

Male sandgrouse take water back to their chicks by carrying it in their feathers. A bird like **BURCHELL'S SANDGROUSE** can transport about two tablespoons of liquid at once.

HIGH TERRAIN

HEAD FOR HIGHER GROUND AND USE IT AS A VANTAGE POINT FROM WHICH TO SURVEY THE HORIZON BELOW TO SPOT ANY POTENTIAL WATER SOURCES. WATER FLOWS DOWNHILL, SO WHEN THE SUN IS LOW IN THE SKY, LOOK FOR THE GLARE OF A REFLECTION ON THE GROUND AS THIS MAY BE A BODY OF WATER.

Vegetation is generally sparse in the desert, but **PALM TREES** grow in areas where water is closer to the surface.

DRAGONFLIES need water to **BREED**. They stay close to water sources, with adult males establishing territories along the edges.

ZEBRAS are constantly on the move looking for new water, and don't often stray more than **7 MILES** from a water source.

FIND YOUR WAY . . . *with a* PUDDLE!

STUDYING RAINFALL IS AN IMPORTANT SKILL FOR ANY NATURAL NAVIGATOR.

PUDDLES CAN TELL US MUCH ABOUT WHAT WILDLIFE AND PEOPLE HAVE BEEN DOING IN AN AREA. THEY CAN ALSO HELP US GET A SENSE OF DIRECTION. WHERE AND HOW WE FIND PUDDLES IN A CITY—LIKE HERE IN MUMBAI, INDIA—HELPS US ORIENT OURSELVES. RAIN RARELY FALLS STRAIGHT DOWN, SO WHERE THE PREVAILING WINDS CARRY THE RAIN IS WHERE PUDDLES WILL FORM. ONCE PUDDLES START TO DRY, THERE ARE CLUES TO SPOT—WHICH SIDE DRIES FIRST AND WHICH ANIMALS USE THEM—THAT CAN HELP US GET OUR BEARINGS, TOO.

With the **SUN** high in the sky in the northern hemisphere, more **PUDDLES** remain on the **SOUTHERN SIDE** of a path.

Can you see any **VIBRATIONS** on the puddle's surface? Although you might not feel any movement, vibrations on a puddle might be a clue that something or someone is approaching.

Puddles are like a **GUEST BOOK**. Water transferred from puddles through **TRACKS** will tell us the direction of travel of a person, vehicle, or animal that paid a visit.

The **GROWTH** on the **SOUTHERN** side of a path will cast a **SHADOW** on the southern part of the path itself.

If there is **BROWN** at one side of a puddle and **GREEN** at the other, then **PLANT MATTER** may have been blown to one end by the **PREVAILING WIND**.

PUDDLE READING

IN THE MIDDLE OF THE DAY IN THE NORTHERN HEMISPHERE, THE SUN IS TO THE SOUTH, WHICH MEANS IT'S PRODUCING SHADOWS TOWARD THE NORTH SIDE OF ANY OBSTACLES. DUE TO THE TEMPERATURE OF THE SUN, IT DRIES ALL MOISTURE EXCEPT THE PARTS IT CAN'T REACH. SO, ON ANY PATH THAT RUNS FROM EAST TO WEST, WE SEE MORE PUDDLES ON THE SOUTHERN SIDE. AS SHADOWS MOVE IN THE OPPOSITE DIRECTION OF THE SUN, IN THE SOUTHERN HEMISPHERE, SHADOWS ARE CAST ON THE SOUTHERN SIDE OF ANY OBSTACLES.

INDIAN PEAFOWL are often found **DRINKING** from puddles. Listen out for their calls (a loud **PIA-OW** or **MAY-AWE**) and follow them to water. The frequency of calling increases before the **MONSOON SEASON**, so it can also be an indicator of incoming **RAIN**.

Feel a **STONE** . . . The **WARMER** side is likely to be the **SOUTHERN SIDE**.

LISTEN *to the* WIND *and* WATCH *the* LEAVES *fall*

FALL SPEAKS TO THE FOREST AND ITS CREATURES.

AS FALL WINDS BLOW THROUGH THE TONGASS NATIONAL FOREST IN ALASKA, THEY TELL ALL WHO TRAVEL THROUGH THAT COOLER DAYS AND FREEZING NIGHTS ARE COMING. SQUIRRELS GATHER AND HIDE NUTS FOR WINTER SNACKS. MOTHER DEER AND THEIR FAWNS GROW THICKER COATS THAT MATCH THE SURROUNDING TREES. BIRDS LEAVE AND HEAD SOUTH. THE COLORFUL LEAVES FALL TO THE GROUND AND OFFER UP A CLEVER WAY FOR YOU TO CONSTRUCT A COMPASS USING THE WIND'S DIRECTION . . .

Leaves have **ORANGE** and **YELLOW** colors, but the chlorophyll that makes their leaves green masks these colors. When cold weather seals off nutrients, the chlorophyll can't be replaced and the leaves gain their **FALL COLORS.**

As the female **SITKA BLACK-TAILED DEER** feels the cold winds on her back, both she and her **FAWN** begin growing **THICKER COATS** to handle the winter cold.

With a **GUST OF WIND,** the fallen leaves **BUMP UP** against **TREES.** Tall trees provide the protection that leaves need from getting pushed around.

The fallen leaves **GATHER IN A PILE** behind the strength of the tree where there is no **PREVAILING WIND.**

THE WAY THE WIND BLOWS

HAVE YOU EVER NOTICED THAT THE WIND BLOWS IN THE SAME DIRECTION MOST OF THE TIME? THAT DIRECTION IS CALLED THE "PREVAILING WIND PATTERN." PREVAILING WINDS ARE WINDS THAT BLOW FROM A SINGLE DIRECTION OVER A SPECIFIC AREA OF THE EARTH. IF YOU KNOW THE DIRECTION OF THE PREVAILING WINDS IN THE AREA YOU ARE IN, YOU CAN USE THAT KNOWLEDGE AND THE NATURE AROUND YOU TO HELP YOU NAVIGATE.

Birds that migrate south for the winter, such as **VARIED THRUSHES**, wait for just the right wind to help them make the journey.

Deciduous trees shed their **LEAVES** in fall to **CONSERVE ENERGY** over the winter, when temperatures and light levels are low, and water may be frozen as ice and snow.

A gust of wind can **DISLODGE NUTS** from trees, giving **SQUIRRELS** a literal windfall of food. If they're not busy eating the nuts, the tree-climbing critters store them before the cold winter arrives.

Pause at a TREE STUMP to DISCOVER ITS PAST

THE INSIDE OF A TREE TELLS ITS STORY.

TREES CAN LIVE FOR HUNDREDS OF YEARS, AND OVER THIS LONG LIFETIME, THEY EXPERIENCE WEATHER SUCH AS RAIN AND SNOW, AND SOMETIMES EVEN FIRE AND FLOODS. WHEN A TREE FALLS OR IS CUT DOWN, THE RINGS LEFT BEHIND IN ITS STUMP CAN TELL YOU NOT ONLY HOW OLD THE TREE IS, BUT WHAT THE WEATHER WAS LIKE DURING EACH YEAR OF ITS LIFE. LOOK CLOSELY AT THE THICKNESS OF THE RINGS OF THIS TREE STUMP HERE IN THE BLACK FOREST, GERMANY, TO DISCOVER ITS PAST LIFE. IT'S THESE RINGS THAT CAN ALSO PROVIDE A SENSE OF DIRECTION.

Coming together for courtship in summer, some **EURASIAN PYGMY OWLS** choose to nest in tree cavities that face away from the **PREVAILING WIND** to protect their future brood.

Follow the **RINGS** of a tree stump toward the middle, where you'll find the **HEART**. This is the point where rings are **SMALLEST**. It tends to be toward the **SOUTHERN EDGE** in the northern hemisphere, and the **NORTHERN EDGE** in the southern hemisphere.

A tree's **BARK** is often **PALER** on the **SOUTHERN SIDE**—the side that receives the most **SUNLIGHT**.

In the northern hemisphere, individual rings will often be **THICKER** on one side due to various environmental conditions like soil nutrition and the direction of the sun. Some trees grow thicker rings on the stump's **SOUTHERN SIDE** since it receives the most **SUNLIGHT**.

A TREE'S STORY

A TREE STUMP DISPLAYS CONCENTRIC GROWTH RINGS. THESE ARE FORMED AS THE TREE GROWS BY CREATING A NEW LAYER OF CELLS. EACH RING MARKS ONE CYCLE OF SEASONS, OR ONE YEAR. IN EACH GROWTH SEASON, TREES CREATE A NEW RING THAT REFLECTS THE WEATHER CONDITIONS OF THAT SEASON. SOMETIMES THE RINGS ARE THICK, WHICH MEANS THAT YEAR WAS FULL OF RAIN. WHEN THE RINGS ARE THIN, THIS MEANS THERE WAS NOT MUCH RAIN THAT YEAR.

The **BARK** of a tree, space between it, and the wood beneath can be home to a huge variety of **INSECTS**, **LICHENS**, and **MOSSES**. These in turn give food to other animals.

WILD BOARS dig for food, and by **SNUFFLING UP THE SOIL**, they create an excellent seedbed for germination. New trees will soon flourish, growing **LONGER, THICKER BRANCHES** on the **SOUTHERN SIDE**.

After **STRONG WINDS** and **HEAVY RAIN**, a tree is saturated with moisture on the side of the prevailing, rain-bearing winds.

ROOTS grow longer and larger on one side of a tree, anchoring it against the **PREVAILING WIND** to prevent it from falling.

Watch a SUNFLOWER *turn to* FACE THE SUN

SUNFLOWERS DANCE UNDER THE SUN.

EACH DAY, YOUNG SUNFLOWERS TRACE THE PATH OF THE SUN ACROSS THE SKY, TURNING THEIR FACES 180 DEGREES FROM EAST TO WEST. THEIR SLOW MOVEMENTS CONTINUE AT NIGHT. AFTER THE SUN SETS, THEY DROOP THEIR TIRED STEMS AND SWING SLOWLY BACK—AWAITING THE RISING SUN TO BEGIN THE CYCLE AGAIN. BY UNDERSTANDING A YOUNG SUNFLOWER'S MOVEMENTS IN RELATION TO THE SUN, YOU CAN LEARN TO ORIENT YOURSELF, BUT THERE ARE A FEW OTHER SECRET SIGNS TO LOOK OUT FOR IN A FIELD OF SUNFLOWERS. TAKE A LOOK AROUND VAL D'ORCIA IN ITALY AND LET NATURE NAVIGATE YOU THROUGH YOUR SURROUNDINGS.

It's late afternoon, and the **YOUNG SUNFLOWERS** are pointing **WEST**.

As a result of the sun's heat, **MATURE SUNFLOWERS** can attract up to five times more **POLLINATORS** because they warm up faster than west-facing plants.

Sunflower gardeners will tell you that a sunflower needs at least **SIX** hours of **DIRECT SUNLIGHT** every day to reach its height potential.

Inside the outer edge of a sunflower sits a bed of up to **TWO THOUSAND** individual flowers—or **FLORETS**. The large array of petals acts as a **NATURAL SIGNPOST** for pollinators and scavengers.

TICK, TOCK, THE INTERNAL CLOCK

IN THEIR BUD PHASE, SUNFLOWERS WILL ALWAYS FACE THE SUN. THIS TRAIT IS CALLED "HELIOTROPISM" AND IS DUE TO A SUNFLOWER'S CIRCADIAN RHYTHM—THE INTERNAL BODY CLOCK THAT HELPS US SLEEP AT NIGHT AND WAKE IN THE MORNING. BUT UNLIKE US, SUNFLOWERS DON'T HAVE MUSCLES, SO HOW ARE THEY TURNING IN THE FIRST PLACE? THAT ANSWER COMES DOWN TO THEIR STEMS. DURING NIGHTTIME, THE STEMS OF YOUNG SUNFLOWERS GROW ON THE WEST SIDE, WHICH ALLOWS THE SUNFLOWER HEAD TO TILT EASTWARD. DURING DAYTIME, THE STEM'S EAST SIDE GROWS, WHICH RESULTS IN IT TURNING TO THE WEST WITH THE SUN.

Once a sunflower **MATURES**, it stops **SUN-TRACKING** and its blooms face **EAST**.

WARM DAYS result in **CHIPMUNKS** surfacing to forage for food. You'll know when they have been in the vicinity as they'll have torn apart a sunflower for **TASTY SEEDS**!

The **BRIGHT GOLDEN** color of the **PETALS** shout, "Hey! Over here!" to seed-eating **BIRDS** such as **GOLDFINCHES**.

Build a MAP from IVY

CLIMBING STEMS AND GLOSSY LEAVES ARE A FAMILIAR SIGHT ON TREES.

IVY IS AN EVERGREEN PLANT AND CAN BE FOUND ALL YEAR ROUND, GROWING MOST VIGOROUSLY IN THE SPRING AND SUMMER. REACHING HEIGHTS OF UP TO 100 FEET, IVY SUPPORTS AT LEAST FIFTY SPECIES OF WILDLIFE, WITH BIRDS AND INSECTS ROOSTING, NESTING, AND HIBERNATING WITHIN IT. AND ITS USEFULNESS DOESN'T STOP THERE! IT IS MOST COMMONLY FOUND AWAY FROM THE DENSE CENTER OF A FOREST, AND YOUNG IVY GROWS IN A DIFFERENT DIRECTION TO MATURE IVY. BOTH THESE THINGS CAN HELP YOU ORIENT YOURSELF WITHIN A WOODLAND, LIKE HERE IN ENGLAND'S NEW FOREST . . .

With its single point and no lobes, **MATURE IVY** grows toward the light from the sun, which in the northern hemipshere is the **SOUTHERN SIDE** of the surface it is growing on.

NEW FOREST PONIES often feed in grassland areas on the **DOWNWIND** side of forested areas to avoid the cold **PREVAILING WIND**.

Ivy is often found growing more abundantly near the edges of **WOODLANDS**.

COMMON IVY, or **ENGLISH IVY**, is not harmful, but in some parts of the world other types of ivy can cause skin rashes. For this reason, touch only those plants you can positively identify.

YOUNG AND MATURE IVY

YOU MAY BE WONDERING HOW WE CAN TELL IF WE'RE LOOKING AT YOUNG IVY THAT'S GROWING AWAY FROM THE LIGHT TO THE NORTH, OR MATURE IVY THAT'S GROWING TOWARD THE LIGHT TO THE SOUTH. DOESN'T ALL IVY LOOK THE SAME? SURPRISINGLY, IT DOESN'T. YOUNG IVY HAS SEVERAL LOBES AND POINTS, WHEREAS MATURE OLDER IVY HAS ONLY ONE POINT.

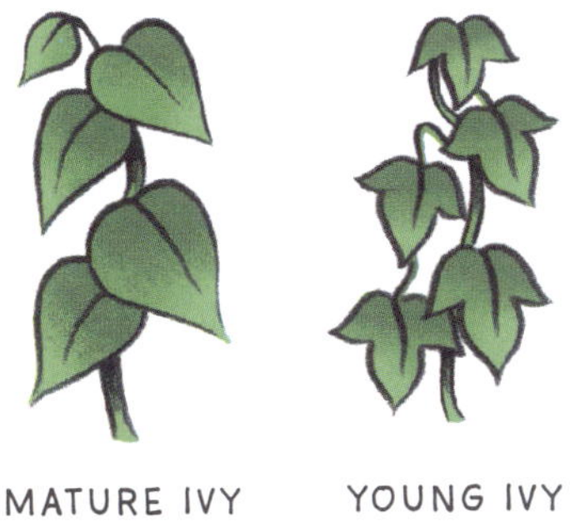

Mature ivy leaves on the **SOUTH** side of trees will **DROOP** slightly so that they get more sun.

Look closely at the vine; you will notice it has many **HAIR-LIKE** roots attaching it to the tree. These roots are on both sides of the vine, but there will be more on the vine's **NORTHERN SIDE**.

With its several lobes, **YOUNG IVY** grows away from the light, toward the **NORTHERN SIDE** of trees.

Discover the secret LANGUAGE *of* MUSHROOMS

LOOK AT, BUT DON'T TOUCH, ONE OF NATURE'S MOST OVERLOOKED SOURCES OF INFORMATION.

AS YOU START TO SEE MORE OF NATURE'S SIGNS, YOU'LL SPOT MUSHROOMS ALL OVER THE PLACE . . . AND NOT JUST IN A TYPICAL WOODLAND, LIKE HERE IN NUUKSIO NATIONAL PARK, FINLAND. YOU WILL DISCOVER DIFFERENT TYPES OF SPECIES IN A GARDEN, IN A HEDGEROW, AND EVEN POKING UP THROUGH THE SIDEWALK! SOME MUSHROOMS LIKE TO GROW IN THE SPRING AND OTHERS IN THE FALL, BUT NEARLY ALL MUSHROOMS LIKE DAMP CONDITIONS. THEY VARY IN COLOR, FROM PINK TO BLACK, AND IN SHAPE, TOO—BUT THEY REMAIN CONSTANT IN THEIR ABILITY TO GIVE US CLUES ABOUT OUR SURROUNDINGS.

The **WOOLLY MILKCAP** grows beneath birch trees in damp areas, mainly on **ACIDIC SOIL**. The distinctive shaggy cap and its relatively unique **PINK COLORING** makes identifying this mushroom very easy.

With its cap looking like a crusty bread roll, it's no surprise the **PENNY BUN**, or **CEP**, is a popular food for the Siberian flying squirrel. The **PENNY BUN** prefers forests with birch trees—particularly **WOODLAND EDGES** and grassy clearings.

Can you smell that? The **DEATH CAP** is **POISONOUS** and releases a **SWEET, SICKLY SMELL** to tell us not to go near it.

You can learn a lot from developing an eye for **MUSHROOM SPECIES**. But if you don't know what type of mushroom it is, don't touch or eat it as it may be **POISONOUS**.

ALL ABOUT MUSHROOMS

MUSHROOMS ARE NEITHER ANIMALS NOR PLANTS BUT BELONG TO THE FUNGI KINGDOM, WHICH DECOMPOSE LIVING OR DEAD PLANTS OR ANIMALS, AND ABSORB THEIR NUTRIENTS. MOST MUSHROOMS GROW FROM TINY SPORES (REPRODUCTIVE CELLS). WHEN SPORES LAND ON THE GROUND, THEY GROW TINY WHITE FIBERS CALLED HYPHAE. THE HYPHAE SPREAD OUT INTO AN UNDERGROUND NETWORK CALLED A MYCELIUM, WHICH HELPS MAKE THE SOIL RICHER. OVER TIME, THE HYPHAE FORM KNOTS THAT PUSH UP THROUGH THE SOIL INTO THE FUNNY-LOOKING FUNGI YOU SEE ABOVE GROUND . . . KNOWN AS MUSHROOMS!

BRACKET-SHAPED MUSHROOMS develop like **SHELVES**, with each row growing below the other. They feed off the wood of the trees they grow on, damaging and even killing them, and are important **NUTRIENT RECYCLERS** in the forest.

MUSHROOMS grow more prominently in places that receive little to no sunlight. In the northern hemisphere, they can more easily be found on the **SHADY, NORTHERN SIDE** of objects like **TREES**.

FLY AGARIC is the mushroom of fairy-tales. It lives in shady spots near **BIRCH TREES**, which are found near the **EDGE OF WOODLANDS**.

CHANTERELLES dot the floors of **BIRCH FORESTS** and mixed hardwood forests. They emit a **FRUITY AROMA**, reminiscent of apricots.

Predict the CHANGING SEASONS by TRACKING migrating animals

SOME ANIMALS MOVE AS THE SEASONS CHANGE, AND WINTER FOLLOWS IN THEIR FOOTSTEPS.

MOST OF THE TIME ANIMALS CAN FIND SHELTER, WATER, AND FOOD WHERE THEY LIVE. BUT AS WINTER ARRIVES, THE WEATHER GETS COLDER, AND FOOD IS HARDER TO FIND, SOME ANIMALS TRAVEL ELSEWHERE. THIS JOURNEY IS CALLED MIGRATION. ANIMALS—LIKE THOSE LIVING HERE IN THE ROCKY MOUNTAINS—CAN JOURNEY SEVERAL THOUSAND MILES, FINDING THEIR WAY OVER LONG AND COMPLEX ROUTES. UNDERSTANDING AN ANIMAL'S MIGRATION PATTERN CAN HELP YOU SPOT THE SIGNS THAT THE SEASONS ABOUT TO TURN.

CANADA GEESE migrate **SOUTH** each winter, flying in a **V-FORMATION** to conserve energy for the long journey ahead.

FRESH SNOW means nothing can pass without leaving a trace. It's easier to track an animal's **FOOTPRINTS** facing into **SUN** as the **SHADOWS** make the impressions stand out.

ELK moving down to the **GRASSY PLAINS** are a sign that the mountain pastures above are becoming snow-covered.

Desperate to escape the cold mountains, **DARK-EYED JUNCOS** migrate **SOUTH**. They're commonly called **SNOWBIRDS** because of their sudden appearance around bird-feeding stations and gardens in winter.

ALTITUDINAL MIGRATION

SOME ANIMALS MAKE LONG JOURNEYS BACK AND FORTH ACROSS LAND AND OCEAN BEFORE WINTER APPROACHES. OTHER MIGRATIONS TAKE A DIRECT ROUTE ACROSS THE SKY. BUT DURING SEASONS OF SEVERE WEATHER IN THE MOUNTAINS, FOR INSTANCE, SOME ANIMALS MAKE REGULAR TRIPS DOWN FROM THE HIGH ALTITUDES INTO THE FOOTHILLS BELOW IN RESPONSE TO CLIMATE AND AVAILABILITY OF FOOD—THIS IS CALLED ALTITUDINAL MIGRATION.

High elevations protect **MOUNTAIN GOATS** from **PREDATORS**, but as winter arrives, they migrate to lower elevations, where they seek **SOUTH-FACING SLOPES.**

Herds of **BIGHORN SHEEP** migrate to lower, warmer grounds during winter. They follow the growth of newly emerged plants, a behavior known as **SURFING THE GREEN WAVE.**

PRAIRIE FALCONS withdraw from the highest-elevation parts of their breeding range in winter, seeking **SHELTER** in low cliffsides.

Watch a BUTTERFLY to see what it is SHOWING YOU

BUTTERFLIES CAN TELL US SO MUCH ABOUT THE WEATHER, THE SEASON, AND EVEN WHERE WE ARE.

WE ALL LOVE BUTTERFLIES FOR THEIR BEAUTIFUL, BRIGHTLY COLORED WINGS—BUT THEY ARE MORE THAN JUST SOMETHING TO ADMIRE. FIRST, TAKE AN INTEREST IN THEIR HABITATS. NEARLY ANYWHERE WITH NECTAR-PRODUCING FLOWERS WILL HOST BUTTERFLIES—FROM DENSE RAINFORESTS TO HOT DESERTS. THEY ARE MOST COMMONLY FOUND WITHIN OPEN, SUNNY GLADES AND ALONG THE EDGE OF WOODLAND PATHS—LIKE HERE IN ENGLAND'S EPPING FOREST—WHERE SOME DEPEND ON STINGING NETTLES FOR THE GROWTH OF THEIR LARVAE. EACH TYPE OF BUTTERFLY HAS DIFFERENT BEHAVIORS . . . DISCOVER WHAT THESE VARIETIES ARE TELLING YOU TODAY!

PEACOCK butterflies are careful about the location of their eggs, preferring the tip of nettles in **FULL SUN**. Ten days later, the black **CATERPILLARS** feed in **SILKEN TENTS** atop the nettles.

STINGING NETTLES thrive in **DISTURBED GROUND**, which means you often come across them on busy, forested **TRAILS**. Once you see them, you are getting closer to **HUMAN HABITATION**.

SMALL TORTOISESHELL butterflies lay **LARGE CLUSTERS** of several hundred green eggs on the underside of fresh nettle leaves between **APRIL** and **MAY**.

Butterflies can give us a **TEMPERATURE CHECK.** As **COLD-BLOODED CRITTERS**, not many can fly in temperatures below **60 DEGREES FAHRENHEIT.**

STINGING NETTLES

MOST BUTTERFLIES LAY THEIR EGGS ON ONLY A FEW SELECT HOST PLANTS. THEY USE THREE OF THEIR SENSES—SIGHT, SMELL, AND TASTE—TO FIND SUITABLE PLANTS, IDENTIFYING CHEMICALS FROM A PLANT THAT INDICATE IT IS A SUITABLE SOURCE OF FOOD FOR THE CATERPILLARS. A COMMON PLANT LIKE THE STINGING NETTLE IS A GREAT PLACE FOR BUTTERFLY LARVAE TO FEED IN LARGE GROUPS IN SILKEN TENTS. THE BUTTERFLIES PICK NETTLE STEMS IN DAPPLED SUNLIGHT, WITH THE STING ALSO FORMING A PROTECTIVE SHIELD AGAINST GRAZING ANIMALS BELOW. SO, IF YOU FIND STINGING NETTLES GROWING ON THE FOREST FLOOR, IT MEANS A BUTTERFLY FAMILY MAY BE NEARBY!

The presence of **ORANGE-TIPS** and **HOLLY BLUES** is a signal that **SPRING** is here. These are two of the first butterfly species to emerge that have spent winter as chrysalises.

RED ADMIRAL butterflies lay a **SINGLE** egg on **UPPER** nettle leaves, choosing leaves that get plenty of **SUNLIGHT.**

Butterflies—like the tiny **COMMON BLUE**—can be easily harmed if there is **RAINFALL.** They seek **SHELTER** under **LARGE** leaves and **BRANCHES** when they sense rain is in the air.

LOCATE *a* SPIDERWEB *and* CATCH THE WIND

A SPIDERWEB SPARKLING WITH MORNING DEW IS A BEAUTIFUL SIGHT.

SOME SPIDERS WEAVE A NEW WEB EVERY NIGHT. USUALLY, A SPIDER'S WEB IS SPUN TO USE AS A HUNTING TOOL WHEN MORNING COMES. BUT ITS SUCCESS DEPENDS ON ITS LOCATION. IN THE AMAZON, BRAZIL, A SPIDER CAN USE THE RAINFOREST'S DENSE VEGETATION TO ITS ADVANTAGE, PICKING A SAFE, SHIELDED SPOT TO SPIN WHERE THE WIND CAN'T REACH IT. SO WHEN WE KNOW THE DIRECTION OF THE PREVAILING WIND, WE CAN SEARCH FOR WEBS TO PROVIDE A HINT AS TO OUR POSITION.

WIND SPEED increases as the height from the ground increases, with maximum wind speed occurring at the **UPPERMOST BRANCHES** of the trees.

Between **JANUARY** and **JULY**, the **PREVAILING WIND** in the Amazon blows from **NORTH** to **SOUTH**, flipping to **SOUTH** to **NORTH** for the last five months of the year.

Many **SPIDERWEBS** are on the **SHELTERED** side of vegetation so they are less exposed to the **WIND**.

The **MORNING DEW** reveals a multitude of webs that were previously invisible because of the transparent nature of silk.

BORN TO SPIN

SPIDERS WEAVE THEIR WEBS BY SQUEEZING A LIQUID FROM THEIR SPINNERETS TO MAKE SILK. THE SILK IS A LIGHT, NATURAL FIBER MADE OF PROTEIN AND IT IS VERY, VERY THIN. A THREAD IN A SPIDERWEB CAN BE A THOUSAND TIMES THINNER THAN A SINGLE HAIR FROM YOUR HEAD! IT CAN BREAK WHEN STRETCHED UP TO TWO TO FOUR TIMES ITS LENGTH. AS SUCH, A SPIDER WILL NOT CONSTRUCT ITS WEB IN PLACES THAT ARE OVEREXPOSED TO THE ELEMENTS, LIKE THE WIND.

Some **JUMPING SPIDERS** will spin tiny tents or shelters to protect themselves from **RAIN** or **PREDATORS**.

GOLIATH BIRDEATERS don't spin webs to trap food. Instead, they use their weaving skills to line their burrows under the forest floor.

Many insects, such as the **RHINOCEROS BEETLE**, thrive on the damp, hot **FOREST FLOOR**, where they are **CONCEALED** from rainfall and sunlight.

Rather than weaving a web at night, **FISHING SPIDERS** take advantage of the wind in the day, standing with their legs up in the air to propel themselves across the **WATER'S SURFACE** toward their prey.

Watch a HUDDLE of SHEEP to forecast the weather

THE HUMBLE SHEEP CAN TEACH US ABOUT NATURE'S ELEMENTS.

IN WALES—LIKE HERE IN THE PRESELI MOUNTAINS—THERE ARE EVEN MORE SHEEP THAN PEOPLE! WE MAY NOT SPEAK THE SAME LANGUAGE AS SHEEP, BUT THESE WOOLLY ANIMALS CAN TELL US A LOT ABOUT THE WEATHER. THEY HAVE TWO PRIMARY GRAZING PERIODS (DURING THE EARLY MORNING AND AGAIN LATE IN THE AFTERNOON) AND ARE KNOWN FOR SPREADING OUT TO GRAZE. IN HOT WEATHER, THEY GRAZE INTO THE WIND, BUT ON COLD, WINDY, AND WET DAYS, THEY GATHER IN TIGHT HUDDLES IN DOWNWIND CORNERS. INCIDENTALLY, DOING SO MAKES IT EASIER FOR THEM TO BE HERDED!

SHEEP are indicators of the **PREVAILING WIND**: They huddle behind **NATURAL RISES** to avoid harsh, cold winds.

SHEEP have a strong **FLOCKING INSTINCT**. They feel safer when they are gathered together.

A ring, or **HALO**, around the sun is caused by refraction of the light through the ice crystals of **CLOUDS** and is a sign we can **EXPECT RAIN** soon.

A HUDDLE OF SHEEP

THERE IS AN OLD RHYME THAT GOES, "WHEN THE SHEEP BEGIN TO HUDDLE, THEN WE'LL SOON HAVE A PUDDLE." SHEEP CANNOT PUT ON RAINCOATS OR POP OPEN AN UMBRELLA LIKE WE CAN, BUT THEY DO HAVE A NATURAL INSULATION TO EXTREME WEATHER IN THE FORM OF THEIR FLEECE. IN COLD, WET, AND WINDY CONDITIONS, SHEEP HUDDLE TOGETHER AND SEEK SHELTER BEHIND WINDBREAKS TO PRODUCE AND CONSERVE HEAT.

Legend has it that when **RAINY WEATHER** is on the horizon, **COWS** can sense increasing air moisture and will lie down to keep a patch of grass dry.

Keep an eye out for any straggly bits of **WOOL**. **SHEEP** leave their mark on any objects they have chosen as their **SHELTER**.

Sheep love to nibble on the small yellow flowers of **GORSE BUSHES**. As a result, bushes often show **HEALTHIER GROWTH** on their **UPWIND SIDE**.

Some **HERDING DOGS** like **COLLIES** have white-tipped tails, known as **LANTERNS**. These tails are easy to see in the diminishing light caused by rain, making it easier to follow them to shelter.

Tune in to the DAWN CHORUS *to* WELCOME THE DAY

THE DAWN CHORUS OCCURS WHEN BIRDS SING AT THE START OF A NEW DAY.

IN EARLY-MORNING HOURS ALL OVER THE WORLD, BIRDS HOLD A SINGING CONTEST. THE BIRDS JOINING IN THE CHORUS VARY BY HABITAT, BUT HERE IN AN ENGLISH GARDEN DURING SPRING, THE ROBIN MAY STRIKE UP WELL BEFORE DAWN. SOON IT IS FOLLOWED BY BLACKBIRDS, WRENS, WARBLERS, AND SPARROWS. OVER THE NEXT HOUR, THE BIRDSONG BUILDS IN WAVES, AND THE GROWING DAWN CHORUS BECOMES A FEATHERED ALARM CLOCK TO TELL US TO WELCOME THE DAY.

BIRDS usually sit **FACING INTO** the wind, so their feathers are not ruffled. But if the wind keeps changing or there is very little wind, they may sit facing different ways.

The first choristers in the **DAWN CHORUS** usually start singing at least an **HOUR BEFORE SUNRISE**. The chorus itself is loudest around **THIRTY MINUTES** on either side of the **RISING OF THE SUN**.

Birdsong is particularly prevalent from **MARCH** to **JULY** in the northern hemisphere, where the **BREEDING SEASON** is during the **WARMEST** part of the year.

A FEATHERED ALARM CLOCK

BIRDS MAY SING AT ANY TIME OF DAY, BUT SONGS ARE OFTEN MORE ENERGETIC, LOUDER, AND MORE FREQUENT IN THE EARLY-MORNING HOURS. THIS CONCERT MAY START AS EARLY AS FOUR A.M. AND EXTEND SEVERAL HOURS UNTIL THE SUN HAS RISEN AND TEMPERATURES BEGIN TO WARM. THIS NATURE'S GET-READY-FOR-THE-DAY SONG IS KNOWN AS THE DAWN CHORUS, AND IS MOST NOTICEABLE IN SPRING WHEN THE BIRDS ARE PERFORMING EITHER TO DEFEND A BREEDING TERRITORY OR ATTRACT A MATE.

The sun travels in an **ARC**, so you can use its light to determine the **CARDINAL DIRECTION**—north, east, south, and west.

Reaching its **HIGHEST POINT** in the sky at midday, the sun points toward **SOUTH** in the northern hemisphere.

Let the SUMMER SUN *illuminate the way*

SINCE THE BEGINNING OF TIME, WE HAVE LOOKED TO THE SUN TO GUIDE US.

EVERY MORNING WITHOUT FAIL THE SUN RISES, BRINGING LIGHT AND WARMTH AT DAYBREAK. IT THEN TRAVELS IN A SMOOTH ARC FROM EAST TO WEST BEFORE IT SETS AT DUSK, CAUSING THE LIGHT TO FADE AND THE AIR TO COOL. BY UNDERSTANDING THE SUN'S PATH THROUGH THE SKY, YOU CAN LEARN TO ORIENT YOURSELF, BUT THERE ARE A FEW OTHER CLUES TO HELP GUIDE YOU, EVEN ON A CLOUDY DAY. TAKE A LOOK AROUND THIS WOODLAND IN BANAGHER GLEN, NORTHERN IRELAND, AND DISCOVER NATURE'S HIDDEN SIGNPOSTS.

Pay attention to the flight of **HONEYBEES**. Honeybees with nectar return to their hive using the sun as a **FIXED REFERENCE POINT**. This allows them to fly in a straight line (or a beeline!) by keeping the angle between their path and the sun constant.

In the northern hemisphere, the most **DENSE LEAVES** of a tree grow on its **SOUTHERN SIDE** because this is where they receive the most **SUNLIGHT**.

A **TREE'S BRANCHES** can grow upward, instead of outward, in the search for sunlight. For trees in an open area in the northern hemisphere, the branches will be more **VERTICAL** on the shaded **NORTHERN SIDE**.

THE SUN'S MOVEMENT

THE SUN ALWAYS TRAVELS IN AN ARC FROM EAST TO WEST, BUT WHERE IT RISES AND SETS CHANGES SLIGHTLY DEPENDING ON THE TIME OF YEAR. THAT IS BECAUSE THE EARTH SPINS ON AN AXIS THAT IS TILTED, CAUSING THE DAYS TO GET LONGER AND SHORTER WITH THE CHANGE IN SEASONS. THIS MEANS THAT IN MIDSUMMER IN NORTHERN IRELAND, IT RISES IN THE NORTHEAST, WHEREAS IN MIDWINTER, IT RISES IN THE SOUTHEAST.

SHADOWS from fallen logs move from **WEST** to **EAST** during the day, in the opposite direction to the passage of the sun.

Jewel-green **MOSS** grows in areas where there is less **LIGHT** and more **MOISTURE**. These simple plants function like sponges, sucking up water and hanging on to it.

EAST AS THE SUN SETS IN THE WEST. THE LIGHT OF A FULL MOON CAN BE SO BRIGHT AS TO CAST ITS GHOSTLY GLOW ON EVERYTHING UNDER ITS GAZE. MANY CREATURES HAVE ADAPTED TO AVOID THE HARSH RAYS OF THE SUN AND LIVE INSTEAD IN SHADOWY MOONLIT MEADOWS, LIKE THESE CREATURES OF THE PYRENEES MOUNTAINS IN SOUTHERN EUROPE. AND JUST AS WE CAN NAVIGATE BY THE SUN IN THE DAY, A FULL MOON CAN HELP US TO ORIENT OURSELVES BY NIGHT, SINCE WE KNOW THAT AT MIDNIGHT IT IS POSITIONED DUE SOUTH.

It's thought that **BATS** hide away during a full moon—some scientists describe their decreased activity at that time as "lunar phobia." This could be because they don't want to be spotted by a potential attacker in the bright light!

Dogs are known to **COMMUNICATE** by **HOWLING** at night. They don't just do this when the moon is full, though. This myth most likely came about because in past times people went outside more often during a full moon, since it offers more light for navigating in the dark.

A PHASE OF THE MOON

THE MOON IS THE EARTH'S ONLY NATURAL SATELLITE AND IT PRODUCES NO VISIBLE LIGHT OF ITS OWN. AS IT ORBITS THE EARTH, YOU SEE THE PARTS OF THE MOON THAT ARE LIT UP BY OTHER OBJECTS—MAINLY THE SUN. WHEN THE MOON IS ON THE OTHER SIDE OF THE EARTH OPPOSITE THE SUN, THEN THE NEAR SIDE OF THE MOON WILL BE FULLY LIT UP. THIS IS CALLED A "FULL MOON" AND IT APPEARS AS A COMPLETE CIRCLE IN THE SKY.

In the northern hemisphere, a full moon will be exactly **SOUTH** in the middle of the night.

The full moon rises and sets like the sun, **RISING IN THE EAST** and **SETTING IN THE WEST**.

You can watch a **FULL MOON'S** progress across the sky and work out the **TIME** in the same way that you would with the sun.

SEEDS absorb the most **WATER** during a full moon, when more moisture is pulled to the **SOIL SURFACE**.

The **HUMMINGBIRD HAWK-MOTH** prefers to fly in sunlight, but it will fly at night if the moon is **BRIGHT**. Its memory brings it to the same **FLOWERBEDS** every day at about the same time.

PHASES OF THE MOON

THERE ARE EIGHT PHASES OF THE MOON: NEW MOON, WAXING CRESCENT, FIRST QUARTER, WAXING GIBBOUS, FULL MOON, WANING GIBBOUS, LAST QUARTER, AND WANING CRESCENT. THE PHASES ARE NAMED AFTER HOW MUCH OF THE MOON WE CAN SEE, AND WHETHER THE AMOUNT VISIBLE IS INCREASING OR DECREASING EACH DAY. THE PHASES WORK IN A CYCLE, AND A COMPLETE CYCLE—FROM NEW MOON TO FULL MOON—TAKES TWENTY-NINE-AND-A-HALF DAYS, KNOWN AS A LUNAR MONTH.

Usually, the **WAXING CRESCENT MOON** rises before noon, being visible in the day sky. Around sunset, it gets even more visible, and it sets before **MIDNIGHT**.

Look up at a WAXING CRESCENT MOON

SEEN FROM EARTH, THE MOON GROWS AND SHRINKS.

THE MOON REVOLVES AROUND THE EARTH, AND AS IT DOES, WE SEE DIFFERENT ANGLES OF ITS ILLUMINATED SURFACE. ON ITS JOURNEY FROM NEW MOON TO FULL MOON, WE SEE A GROWING PORTION OF ITS ILLUMINATED HALF. THIS IS CALLED A "WAXING MOON," WHICH APPEARS BIGGER IN THE SKY EVERY NIGHT. AFTER FIFTEEN DAYS, IT BECOMES A FULL MOON. THEN, THE PROCESS REVERSES AND ITS ILLUMINATED PORTIONS SHRINK EVERY NIGHT. THIS IS CALLED A "WANING MOON." IN THE NORTHERN HEMISPHERE—LIKE HERE IN THE ARCTIC—WE SEE THE WAXING CRESCENT PHASE AS A THIN CRESCENT OF LIGHT ON THE RIGHT. IN THE ARCTIC, THE TIMEFRAMES AND OBSERVATIONS WILL BE SLIGHTLY DIFFERENT DUE AN OBSERVER'S PROXIMITY TO THE NORTH POLE. LET'S LOOK UP AND INTERPRET THE SIGNS THIS SLIM SMILE OF A MOON CONVEYS.

ARCTIC REINDEER find vegetation by moving snow around with their antlers. While feeding, some may put their **TAIL TOWARD** the **PREVAILING WIND** to keep the cold air off their face and chest.

The two points of a **CRESCENT MOON**, also known as their horns, aim directly **AWAY** from the **SUNSET**.

When observing the moon in the northern hemisphere, a quick glance will let you know if it's waxing. When you can draw a **"P"** with an **IMAGINARY LINE** down the **LUMINOUS CRESCENT**, it's a **WAXING MOON**.

Draw an imaginary line that **CONNECTS** a **CRESCENT MOON'S** horns. **CONTINUE** the line down to the horizon where you'll find **SOUTH**.

Imagine the crescent as an **ARCHER'S BOW**: the invisible arrow is aimed directly at the sun, which is **BELOW THE HORIZON**.

Jagged snow and ice ridges, called **SASTRUGI**, are sculpted by the **PREVAILING WIND**. Their sharp edges align with the wind and remain **CONSTANT IN DIRECTION** for as long as the wind blows.

Some **ARCTIC FOXES** create their homes in caves that **FACE AWAY** from **PREVAILING WINDS** to protect their cubs from the harsh cold.

Use the NORTH STAR *as your guide*

THE STARS CAN ACT AS OUR BEACONS.

JUST LIKE THE SUN, STARS TRAVEL ACROSS THE SKY FROM EAST TO WEST. BY KEEPING TRACK OF WHICH DIRECTION THEY SEEM TO BE MOVING, YOU CAN DETERMINE WHICH WAY YOU'RE FACING. IN THE NORTHERN HEMISPHERE, LIKE HERE ON THE CABOT TRAIL IN NOVA SCOTIA, CANADA, THERE IS ONE STAR IN THE NIGHT SKY THAT DOES NOT APPEAR TO MOVE. IT IS CALLED POLARIS, ALSO KNOWN AS THE NORTH STAR. IT ISN'T THE BRIGHTEST STAR IN THE SKY, AS IS COMMONLY BELIEVED, BUT A MODERATELY BRIGHT SECOND-MAGNITUDE STAR. EVEN SO, IT'S STILL STRONG ENOUGH TO BE SEEN IN A DARK SKY. POLARIS SITS DIRECTLY OVER THE EARTH'S NORTH POLE AND WE CAN USE IT TO FIND TRUE NORTH . . .

Once you have determined which direction is north, you can use the North Star to work out how far north you are—your **LATITUDE**. Facing the North Star (so, looking north), it will be the **SAME ANGLE** above the **HORIZON** as your latitude.

Under the glow of the stars, **AMERICAN MARTENS** move deftly among the trees. They use **SCENT GLANDS** on their abdomen to mark their route through the treetops, allowing them to travel faster in search of prey.

LUNA MOTHS are a favorite snack for nocturnal **BATS**. To protect themselves, the large green insects flutter their **WING TAILS**, causing the bats to zero in on the lower wings rather than more vital body parts.

If you follow this line for about **FIVE TIMES** the distance between Dubhe and Merak, the next bright star you reach will be the **NORTH STAR**.

NORTH STAR

Look up . . . These seven stars are the **BIG DIPPER**, which is part of the Ursa Major constellation. Use the **TWO STARS** (known as Merak and Dubhe) at the end of the **BIG DIPPER** to indicate a direction.

DUBHE

MERAK

Looking at the North Star, if you imagine a line **VERTICALLY** down from the North Star to the **HORIZON**, this is **NORTH**. **EAST** is to the **RIGHT** and **WEST** to the **LEFT**. **SOUTH** is **BEHIND** you.

The constellation **CASSIOPEIA** is helpful in finding the North Star, as it will always be on the **OPPOSITE SIDE** of the North Star from the Big Dipper and therefore often high in the sky.

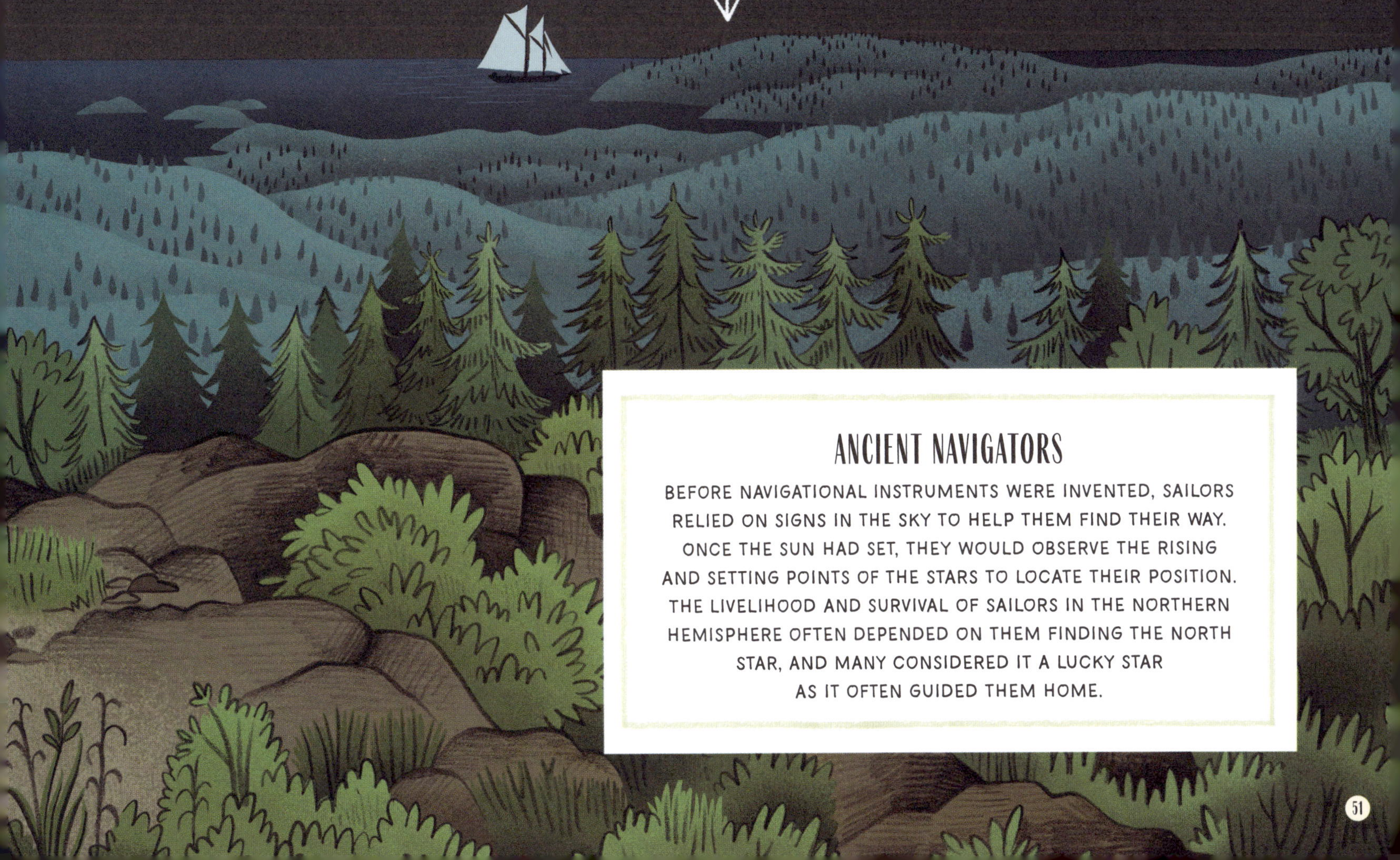

ANCIENT NAVIGATORS

BEFORE NAVIGATIONAL INSTRUMENTS WERE INVENTED, SAILORS RELIED ON SIGNS IN THE SKY TO HELP THEM FIND THEIR WAY. ONCE THE SUN HAD SET, THEY WOULD OBSERVE THE RISING AND SETTING POINTS OF THE STARS TO LOCATE THEIR POSITION. THE LIVELIHOOD AND SURVIVAL OF SAILORS IN THE NORTHERN HEMISPHERE OFTEN DEPENDED ON THEM FINDING THE NORTH STAR, AND MANY CONSIDERED IT A LUCKY STAR AS IT OFTEN GUIDED THEM HOME.

Locate the SOUTHERN CROSS

SOUTHERN CROSS

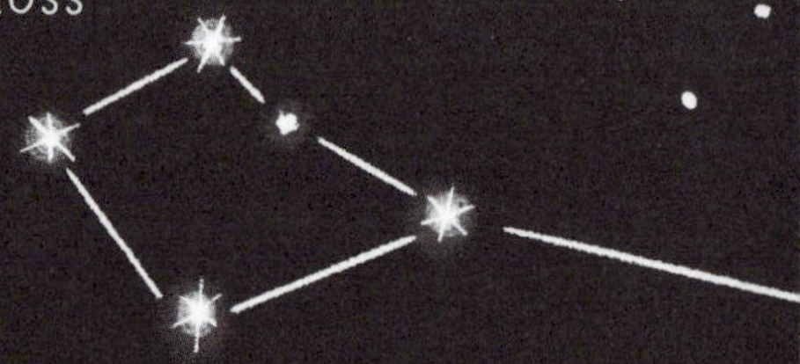

Look up... **LOCATE** the **FIVE-STAR CLUSTER** known as the **SOUTHERN CROSS**.

ON A BRIGHT CLEAR NIGHT, WHAT DO YOU SEE WHEN YOU LOOK UP AT THE SKY?

MANY PEOPLE SAY THAT THE STARS APPEAR AS A SHINING BLANKET, TWINKLING OVERHEAD, BUT STARS CAN TELL US A LOT ABOUT WHERE WE ARE. THE SOUTH CELESTIAL POLE IS THE POINT IN THE SKY DIRECTLY ABOVE EARTH'S SOUTHERN AXIS. UNLIKE THE SKY OVER THE NORTH POLE, WHICH HAS THE BRIGHT NORTH STAR SHOWING US WHERE TO GO, THERE IS NO SUCH STAR OVER THE SOUTH POLE. HOWEVER, IF YOU ARE IN THE SOUTHERN HEMISPHERE – LIKE HERE IN AUSTRALIA'S NORTHERN TERRITORY – YOU CAN USE THE SOUTHERN CROSS, ALSO KNOWN AS THE CRUX CONSTELLATION, TO FIND YOUR WAY...

POINTERS

Below the Southern Cross, you'll find **TWO BRIGHT STARS** also known as the **POINTERS**.

THE SOUTHERN CROSS

WHEN YOU HEAD OUTSIDE TO STARGAZE, WHAT YOU WILL NOTICE FIRST ARE TWO BRIGHT STARS ON A DIAGONAL. THE BRIGHTER OF THE TWO IS CLOSER TO THE HORIZON. TOGETHER, THESE TWO BRIGHT STARS ARE COMMONLY KNOWN AS THE 'POINTERS'. THAT'S BECAUSE THEY SEEM TO POINT TOWARDS THE SOUTHERN CROSS. THE SOUTHERN CROSS ITSELF IS ABOVE AND A LITTLE TO THE LEFT OF THE POINTERS; IT'S A COMPACT GROUP OF FIVE BRIGHT STARS AND LOOKS LIKE A DIAMOND LYING ON ITS SIDE.

DINGOES mainly hunt at **DAWN**, **DUSK** and **DURING THE NIGHT**. They are found across most of Australia, from grasslands to rainforests, though they favor **EDGES** of forests next to grasslands.

Take the **DIAGONAL** of the box-shape that the **SOUTHERN CROSS** makes and follow an **IMAGINARY LINE** across the sky. Find the middle of the **POINTERS** and do the same.

You'll find the **SOUTH CELESTIAL POLE** where the two lines intersect.

SOUTH CELESTIAL POLE

Look down... from the south celestial pole to the **HORIZON**. You've found **SOUTH**!

SUGAR GLIDERS can be found in star-dappled **EUCALYPTUS TREES** searching for scrumptious sap. As **NOCTURNAL** animals, they see well in the dark with their big black eyes.

Keep discovering

NATURE'S SIGNS . . . *everyday!*

NATURE CHANGES ALL THE TIME, SO THERE'S ALWAYS SOMETHING NEW TO DISCOVER ON YOUR ADVENTURES OUTDOORS. TUNE IN TO YOUR SENSES WHEREVER YOU ARE, AND REMEMBER THAT NATURE'S SIGNS ARE NOT A SECRET TO THOSE WHO LOOK FOR THEM . . .

BE PRESENT

BECOMING A NATURAL NAVIGATOR IS MOSTLY A MATTER OF BEING ATTENTIVE. IT'S A HABIT THAT YOU CAN DEVELOP, BUT IT TAKES PRACTICE. FOCUS ON THE HERE AND NOW, AND SEE WHAT NATURE IS REVEALING TO YOU . . .

BECOME FAMILIAR

LEARN ABOUT WHAT KIND OF HABITATS ANIMALS AND PLANTS PREFER, WHEN THEY BREED, WHAT THEY EAT, AND HOW LONG THEY LIVE. THE MORE NATURAL HISTORY KNOWLEDGE YOU HAVE, THE MORE YOU WILL SEE AND LEARN.

STUDY OBSERVATIONS

THE LONGER YOU CONTINUE LOOKING AT NATURE, THE MORE VALUABLE YOUR OBSERVATIONS WILL BECOME. AND THE MORE YOU THINK ABOUT THE PATTERNS, THE MORE PROFOUND YOUR UNDERSTANDING OF NATURE WILL BE.

ASK QUESTIONS

ONCE YOU CAN IDENTIFY A PLANT OR ANIMAL, DON'T STOP THERE. WITH TIME, YOU WILL BE ABLE TO ASK MORE COMPLICATED QUESTIONS AND ANTICIPATE WHEN DIFFERENT EVENTS IN NATURE WILL NEXT OCCUR.

TEACH OTHERS

ONE OF THE BEST WAYS TO SOLIDIFY WHAT YOU KNOW ABOUT NATURE IS TO SHARE YOUR KNOWLEDGE WITH OTHERS. IF YOU KNOW SOMETHING WELL ENOUGH TO BE ABLE TO EXPLAIN IT, THEN YOU TRULY UNDERSTAND IT.

GLOSSARY

ACID SOIL Soil with a pH of 6 or lower; found especially in cool, moist areas where soluble bases are washed away.

ACTINOMYCETES A group of bacteria that form symbiotic relationships with many types of plants. Together with the plant, the actinomycete fixes nitrogen in the soil, which the plants use to live, enabling the bacterium to continue to thrive.

AIR PRESSURE The force of air pressing down on an area.

ALGAE A type of plant with no stems or leaves that grows in water or on damp surfaces.

ALTITUDINAL MIGRATION A short-distance animal migration from lower altitudes to higher altitudes and back.

ANVIL CLOUD Another name for a cumulonimbus cloud, which tends to be anvil-shaped (a flat top and concave sides).

ATMOSPHERIC PRESSURE The pressure of the atmosphere on the earth's surface.

AXIS A line through the center of an object, around which the object turns. Earth makes a complete turn on its axis every twenty-four hours.

BACTERIA Small organisms, or living things, that can be found in all natural environments.

BREEDING The mating and production of offspring by animals.

CAMOUFLAGE The way in which some animals are colored and shaped so that they cannot easily be seen in their natural surroundings.

CARDINAL DIRECTIONS The four main points of a compass: north, south, east, and west.

CASSIOPEIA A large constellation located in the northern sky.

CAT'S-PAW A pattern of ripples on the surface of water caused by a light wind.

CAVITY A hollow space, like a hole in a tree.

CHLOROPHYLL A green substance in plants that enables them to use the energy from sunlight in order to grow.

CIRCADIAN RHYTHM An internal process that regulates the sleep—wake cycle and repeats roughly every twenty-four hours.

CLOUD A mass of condensed or visible water vapor that floats in the sky.

CLUSTER A number of things occurring close together.

COLD-BLOODED Animals with a body temperature that changes according to the surrounding temperature.

COLONIZE When plants and animals become established in a new environment.

COMPASS An instrument used for finding direction.

CONCENTRIC Circles or rings that have the same center.

CRESCENT MOON The moon seen in the shape of a crescent when waxing (growing) or waning (shrinking).

CUMULONIMBUS A type of cloud that extends to a great height and is associated with thunderstorms. See also "anvil cloud."

DAWN CHORUS The singing of birds at dawn.

DECIDUOUS A tree or bush that loses its leaves in the fall every year.

DECOMPOSER Any organism in a community, such as a bacterium or fungus, that breaks down dead tissue, enabling the constituents to be recycled to the environment.

DEW Drops of water that form on the ground and other surfaces outdoors during the night.

DOWNWIND The direction in which the wind is blowing.

DRIFT SNOW A bank of deep snow heaped up by the wind.

DUNE A mound or ridge of drifted sand.

EARTH The planet on which we live.

EVERGREEN Having green leaves throughout the entire year.

FAUNA All the animal life in a particular place.

FLOCK A group of animals of one kind, especially sheep or birds.

FLORA All the plant life in a particular place.

FLORET A small flower, one of many making up the head of a composite flower.

FORAGE To travel from place to place searching for things that you can eat or use.

FORECAST To predict or calculate in advance, like the weather.

FOREST FLOOR The above-ground layer of a forest made up of tree roots, soil, and decaying matter.

FULL MOON A phase of the moon occurring when the Earth lies between the sun and the moon so that the moon is visible as a fully illuminated disc.

FUNGI Any of a group of spore-producing organisms feeding on organic matter, including mushrooms and toadstools.

GIBBOUS Having the illuminated part of the moon greater than a semicircle and less than a circle.

HABITAT The natural home or environment of an animal, plant, or other organism.

HALO A circle of light around a person or thing, like the sun.

HELIOTROPISM The growth of plants in response to the stimulus of sunlight, so that they turn to face the sun.

HIBERNATE When animals or plants spend the winter season in a dormant or inactive state.

HIGH PRESSURE Atmospheric pressure that is higher than normal.

HIGH TERRAIN Being a relatively great distance above the ground or above sea level.

HORIZON The line at which the Earth's surface and the sky appear to meet.

HOST PLANT A plant upon which an organism, such as an insect, lodges and subsists.

HUMIDITY The amount of water in the air.

HYPHAE Each of the branching filaments that make up the mycelium of a fungus.

ICEBLINK A bright reflection of sunlight on the bottom of a low cloud, caused by ice on a distant expanse of water or land.

INVERTEBRATE A creature that does not have a spine; for example, an insect or a worm.

KELP A type of flat brown seaweed.

LARVAE An immature, free-living form of many animals that develops into a different adult form by metamorphosis.

LATITUDE The angular distance of a place north or south of the Earth's equator, or of the equator of a celestial object, usually conveyed in degrees and minutes.

LEEWARD In the direction toward which the wind blows.

MICROSCOPIC Things that are very small and usually can be seen only through a microscope.

MIDSUMMER The period in the middle of the summer.

MIGRATION The seasonal movement of animals from one region to another.

MONSOON The season in southern Asia when there is a lot of very heavy rain.

MOON The natural satellite of the Earth, visible at night by reflected light from the sun.

MOSS A very small, soft green plant that grows on damp soil, or on wood or stone.

MYCELIUM The vegetative part of a fungus or fungus-like bacterial colony, consisting of a mass of branching, thread-like hyphae.

NECTAR A sweet liquid produced by flowers, which bees and other insects collect.

NEST A home that a group of insects or other creatures make to live in and give birth to their young.

NOCTURNAL Active at night.

NORTHERN HEMISPHERE The half of the Earth that is north of the equator.

NORTH POLE The point on the surface of the Earth that is furthest north.

NORTH STAR The star of the northern hemisphere toward which the axis of the Earth points.

ORBIT The curved path in space that is followed by an object going round a planet, moon, or star.

ORIENT To align or position something relative to the points of a compass.

PETRICHOR A sweet smell that is produced when rain falls on dehydrated earth.

PHOTOSYNTHESIS The way that green plants make their food using sunlight.

POLLINATOR An animal that transfers pollen from the male anther of a flower to the female stigma of a flower.

PREDATOR An animal that kills and eats other animals.

PREVAILING WIND A wind that blows predominately from a single general direction.

PUDDLE A small, shallow pool of liquid that has spread on the ground.

RAINBOW A bow-shaped display in the sky of the colors of the light spectrum, caused by the refraction and reflection of the sun's rays through rain.

REFLECTION The production of an image by, or as if by, a mirror.

ROOST A place where birds or bats rest or sleep.

ROOTS The part of a plant that attaches it to the ground or acts as a support, typically underground, and that conveys water and nourishment to the rest of the plant via numerous branches and fibers.

SASTRUGI Long, wave-like ridges of hard snow, formed vertically to the direction of the wind.

SCAVENGER An animal or other organism that feeds on dead organic matter.

SEASON Each of the four divisions of the year (spring, summer, fall, and winter), marked by particular weather patterns and daylight hours, and resulting from the Earth's changing position with regard to the sun.

SEAWEED Any of the red, green, or brown marine algae that grows in the sea or on land very close to the sea.

SHADOW A dark shape on a surface that is made when something stands between a light and the surface.

SNOW Soft white flakes of ice that fall from the sky to the ground.

SOUTH CELESTIAL POLE The point in the sky about which all the stars seen from the southern hemisphere rotate.

SOUTHERN HEMISPHERE The half of the Earth that is south of the equator.

SOUTH POLE The point on the surface of the Earth that is furthest south.

SPECIES A group of organisms whose members have the same main characteristics and are able to breed with one another.

SPORES Cells produced by bacteria and fungi that can develop into new bacteria or fungi.

STEM The thin, upright part of a plant on which the flowers and leaves grow.

STUMP The part of a tree that remains in the ground after the tree falls or is cut down.

SUN The star around which the Earth orbits.

SUNRISE The time in the morning when the sun first appears in the sky.

SUNSET The time in the evening when the sun disappears out of sight from the sky.

SWELL Waves that move up and down in the open sea.

THUNDERSTORM A storm in which there is thunder, lightning, and a lot of heavy rain.

TRACKS Marks, such as footprints, left by someone or something that has passed by.

TREE RING Each of a number of concentric rings in the cross section of a tree trunk, representing a single year's growth.

UPWIND In the opposite direction to that in which the wind is blowing.

VARIABLE WINDS Winds that blow in a small area and are related to air-pressure systems.

VEGETATION Plants, especially those found in a particular area or habitat.

VINE A plant that grows up or over things.

WAVE A long body of water curling into an arched form and breaking on the shore.

WINDWARD The side of something that is facing the wind.

WIND WAVES Waves that are generated by the transmission of momentum and energy from the air to the water.

FURTHER READING

Find out more about natural navigation with the help of an adult.

AMERICAN MUSEUM OF NATURAL HISTORY

AMNH.ORG

The American Museum of Natural History is a fantastic hub to learn about the natural sciences.

NASA KIDS' CLUB

NASA.GOV/KIDSCLUB

Discover more about Earth, the solar system, and beyond.

NATIONAL GEOGRAPHIC KIDS

KIDS.NATIONALGEOGRAPHIC.COM

Explore your world with educational quizzes and videos.

NATIONAL PARK SERVICE'S JUNIOR RANGERS

NPS.GOV/KIDS/JUNIOR-RANGER-ONLINE.HTM

Get to know America's national parks with virtual tours, activities, and more.

To all people of the past and present who have paid attention to their surroundings and shared that information with others —C.C.

For Mom and Dad, my first nature guides —C.S.

The illustrations for this book were created digitally and with graphite.
Set in Above the Sky, Bakerie, Farmhand and Hind.

Library of Congress Control Number: 2021945013
ISBN 978-1-4197-5927-7

Foreword by Steve Backshall
Book design by Nicola Price

First published in the United Kingdom in 2022 by Magic Cat Publishing Ltd.
First published in North America in 2022 by Magic Cat Publishing, an imprint of ABRAMS.

Printed and bound in China
10 9 8 7 6 5 4 3 2

ABRAMS The Art of Books
195 Broadway, New York, NY 10007
abramsbooks.com